SEP 08 2021

Ocean Animals
Jellyfish

by Derek Zobel

BELLWETHER MEDIA
MINNEAPOLIS, MN

Blastoff! Beginners are developed by literacy experts and educators to meet the needs of early readers. These engaging informational texts support young children as they begin reading about their world. Through simple language and high frequency words paired with crisp, colorful photos, Blastoff! Beginners launch young readers into the universe of independent reading.

Sight Words in This Book

a	can	how	not	they
and	do	in	of	this
are	eat	is	out	time
be	from	it	the	to
big	has	look	their	water
called	have	many	there	

This edition first published in 2021 by Bellwether Media, Inc.

No part of this publication may be reproduced in whole or in part without written permission of the publisher. For information regarding permission, write to Bellwether Media, Inc., Attention: Permissions Department, 6012 Blue Circle Drive, Minnetonka, MN 55343.

Library of Congress Cataloging-in-Publication Data

Names: Zobel, Derek, 1983- author.
Title: Jellyfish / by Derek Zobel.
Description: Minneapolis, MN : Bellwether Media, Inc., 2021. | Series: Ocean animals | Includes bibliographical references and index. | Audience: Grades PreK-2
Identifiers: LCCN 2020007748 (print) | LCCN 2020007749 (ebook) | ISBN 9781644873250 (library binding) | ISBN 9781681038124 (paperback) | ISBN 9781681037882 (ebook)
Subjects: LCSH: Jellyfishes--Juvenile literature.
Classification: LCC QL377.S4 Z63 2021 (print) | LCC QL377.S4 (ebook) | DDC 593.5/3--dc23
LC record available at https://lccn.loc.gov/2020007748
LC ebook record available at https://lccn.loc.gov/2020007749

Text copyright © 2021 by Bellwether Media, Inc. BLASTOFF! BEGINNERS and associated logos are trademarks and/or registered trademarks of Bellwether Media, Inc.

Editor: Amy McDonald   Designer: Andrea Schneider

Printed in the United States of America, North Mankato, MN.

Table of Contents

Jellyfish!	4
Body Parts	10
Sting!	18
Jellyfish Facts	22
Glossary	23
To Learn More	24
Index	24

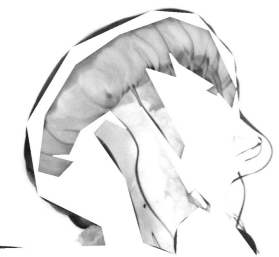

Jellyfish!

Look in the water!
A jellyfish!

Jellyfish are also called jellies. They are not fish!

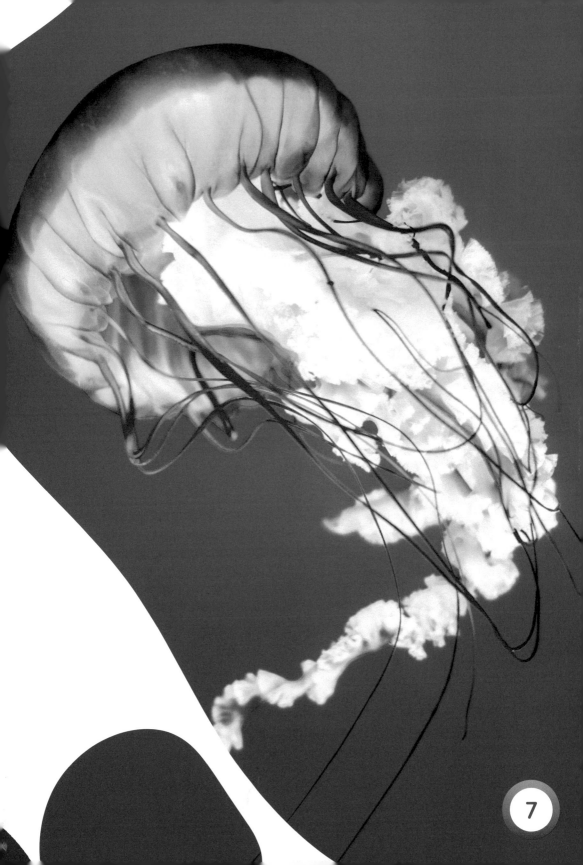

There are many kinds of jellies. They can be big or small.

moon

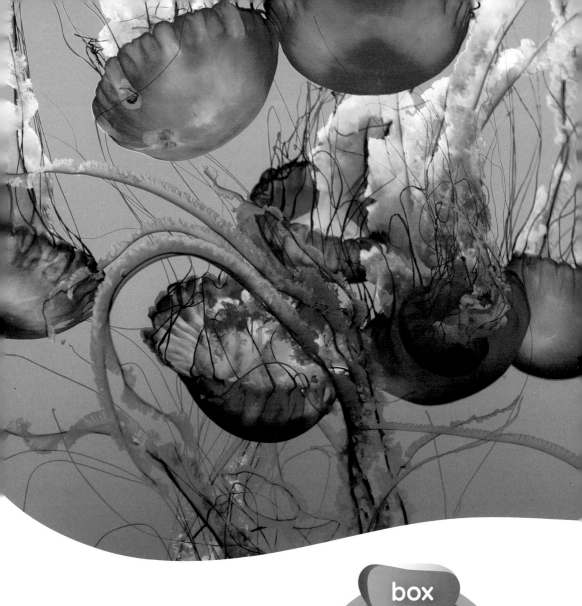

lion's mane

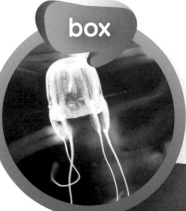

box

Body Parts

Jellyfish bodies are called **bells**. They do not have bones.

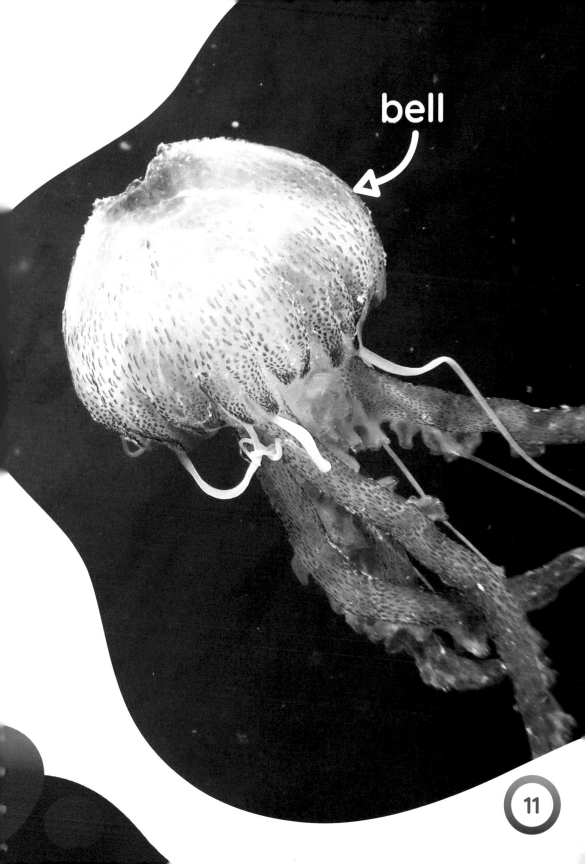

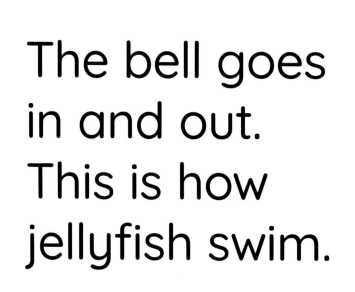

The bell goes in and out. This is how jellyfish swim.

The bell has
a mouth.
It is in the middle.

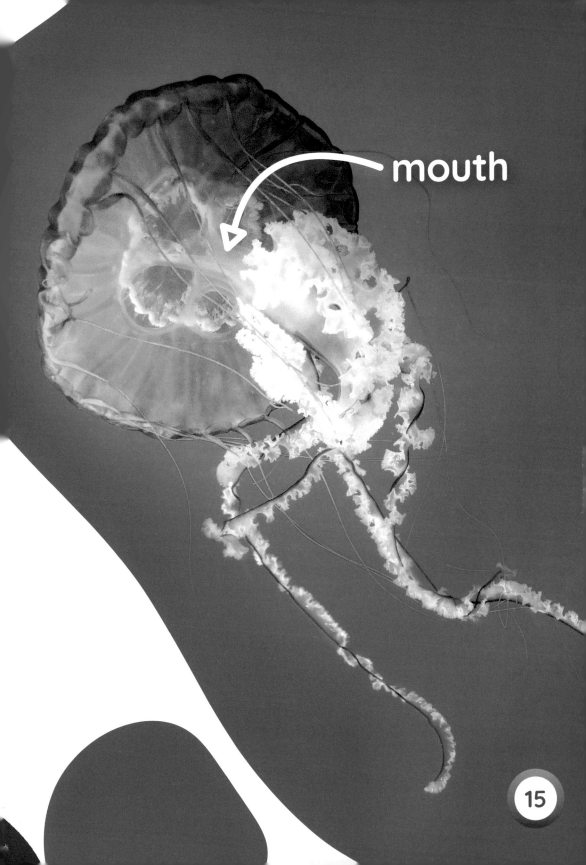

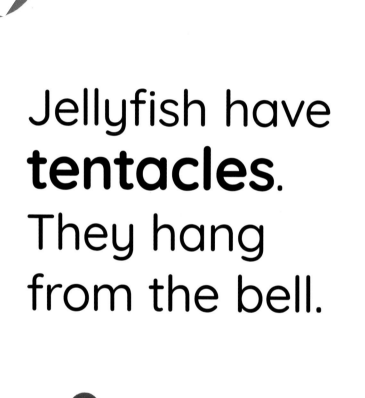

Jellyfish have **tentacles**. They hang from the bell.

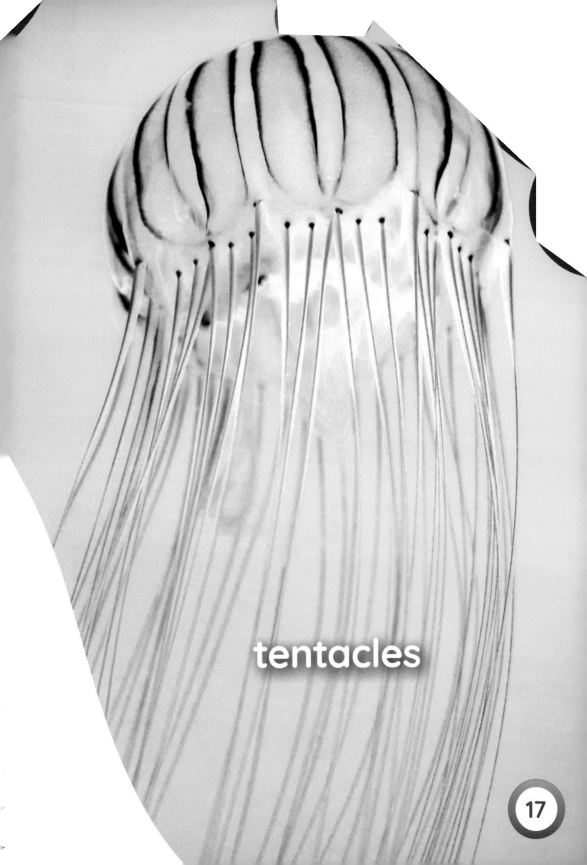

sting!

Tentacles **sting**. Their **venom** stops food.

This jellyfish has a fish. Time to eat!

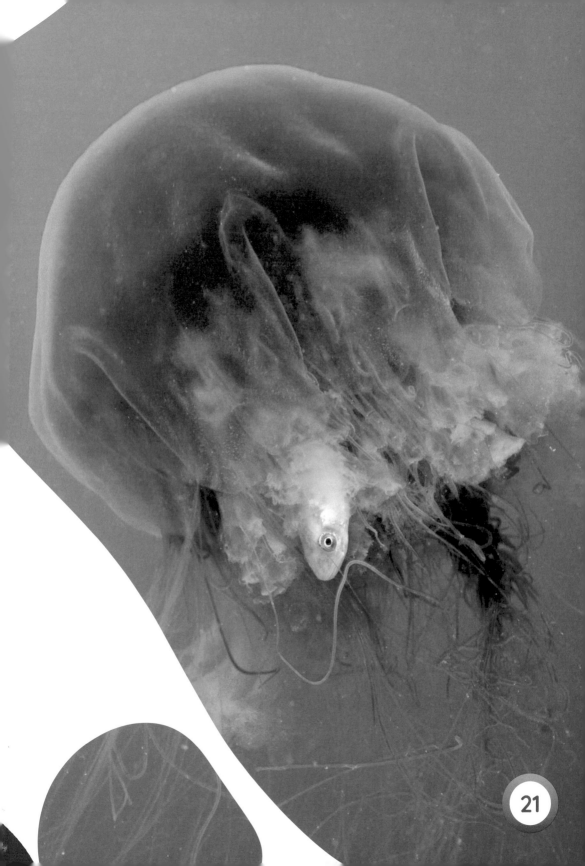

Jellyfish Body Parts

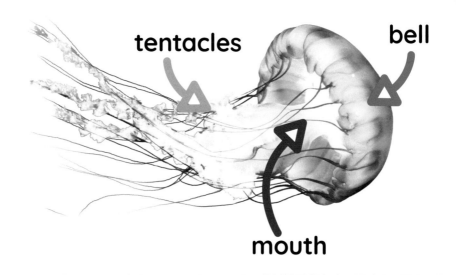

tentacles

bell

mouth

Jellyfish Food

fish   shrimp   small plants

Glossary

bells — the bodies of jellyfish

sting — to poke and use venom

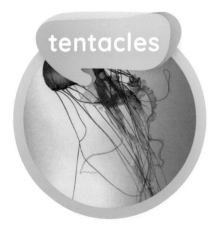

tentacles — long body parts that can sting

venom — a toxin that stops animals

To Learn More

ON THE WEB

FACTSURFER

Factsurfer.com gives you a safe, fun way to find more information.

1. Go to www.factsurfer.com.

2. Enter "jellyfish" into the search box and click 🔍.

3. Select your book cover to see a list of related content.

Index

bells, 10, 11, 12, 14, 16
bodies, 10
bones, 10
fish, 6, 20
food, 18
kinds, 8
mouth, 14, 15
sting, 18

swim, 12
tentacles, 16, 17, 18
venom, 18
water, 4

The images in this book are reproduced through the courtesy of: Beautiful landscape, front cover, pp. 3, 22 (parts); Pawel Kalisinski/ Pexels, pp. 4-5; LeeYiuTung, pp. 6-7; Cassiohabib, pp. 8-9; Laura Dinraths, p. 8 (moon); RLS Photo, p. 9 (lion's mane); TheGift777, p. 9 (box); Tiago Sa Brito, pp. 10-11; Chai Seamaker, pp. 12-13; Norman Chan, pp. 14-15; Ultima_Gaina, pp. 16-17; WaterFrame/ Alamy Stock Photo, pp. 18-19; Nature Picture Library/ Alamy Stock Photo, pp. 20-21, 23 (venom); eventravels, p. 22 (fish); MF Choi, p. 22 (shrimp); Davdeka, p. 22 (small plants); Steve agreca, p. 23 (bells); DiveSpin.Com, p. 23 (sting); qingyi, p. 23 (tentacles).